THE ALIGNMENT PRESCRIPTION

The Simple Remedy For Empowering Technology Organizations To Cure Cancer

BRAD FRANCIS

Foreword By Anthony Goonetilleke

Published By Executech Solutions LLC

Print ISBN: 978-1-7375163-0-9
eBook ISBN: 978-1-7375163-1-6
Audio ISBN: 978-1-7375163-2-3

092021

DEDICATION

This book is dedicated to My Lovely Wife, Cherie, to whom I have been blessed to share decades of love and laughter sprinkled with a bundle of successes, a few challenges, and countless memorable milestones. As the Tavares song goes, "Heaven Must Be Missing an Angel" because she is here with me right now.

“Winning is about having the
whole team on the same page.”

— Bill Walton

PROMISES

The Alignment Prescription is my simple "CIO Rx" remedy for empowering a technology organization to cure the alignment challenge, achieve consistent positive results, and win big! I wrote this book as a quick read that would not bog you down in the muck. To that point, my first promise is that with a commitment in the range of a couple of hours of dedicated reading time, you will be able to learn about the remedy and perform a quick assessment on its fit for you and your organization.

Many authors get caught up in theories that provide varied ideas and alternatives that may be considered. I am not one of those authors and this is not one of those books. The remedy being presented has been production verified; it works. *The Alignment Prescription* promises to focus solely on a proven solution for creating internal organizational alignment, it is a winning formula. I will not leave you with the need to perform a complex analysis as there is an opportunity to succeed based solely on what you read.

You may feel I'm overstating things with the idea that you will be able to empower technical organizations to fight a vicious disease, but the inability to achieve consistent positive results is an organizational cancer you must cure. Remedy internal alignment and you will enable your teams to achieve greatness. Fail to get results and technical leaders will undoubtedly find themselves on business life support. Let us be honest, the demands on senior technology leaders continue to grow, consistent with a supply of projects that is seemingly never ending. You cannot stop that flow, but you can more effectively direct it. This book will provide a solution for synchronizing demand in a manner that enlightens others regarding the impact of their requests. This alignment remedy promises to build organizational awareness, ensure individual and team accountability, and drive the achievement of consistent positive outcomes. No doubt about it, results matter, winning is required, and this is the shot in the arm that technology teams need.

Finally, after reading this book if you are interested in implementing this remedy in your organization, I promise to respond to your questions and provide input where appropriate. Should you express interest, I will also provide information on an automated software solution for your consideration, ultimately allowing you to roll out the approach more efficiently to leverage its benefits. It will quickly get your entire organization on the same page. That said, using our software solution is not a requirement for your success, it is an option.

The bottom line is that *The Alignment Prescription* promises to be a quick read that presents a proven solution for generating awareness, accountability, and achievement in your organization. As a senior technology leader, if you promise to read the book to conclusion, you will understand the significance of this WIN perspective!

CiO

Rx

CONTENTS

"If everyone is moving forward together,
then success takes care of itself."

— Henry Ford

FOREWORD

By ANTHONY GOONETILLEKE

As with most technology executives, throughout the course of my career I have had countless discussions and meetings driven by the need to accelerate, invest, transform, and around achieving business results in an increasingly competitive environment. Whether I am on stage at an industry conference, hosting a company meeting or being asked to speak at an event, the recurring theme always comes back to a focus on consistent improvement. In a world of accelerated technology innovation coupled with the constant disruption of traditional business models, the art of building great teams and executing on a technology vision becomes paramount to the success of every business and its strategy.

The status quo is no longer an option in today's environment, if it ever was, but consistent and quickened advancement is a requirement that is here to stay. My technology journey has been a progression that has included working with terrific technology executives representing a variety of industries. I had the privilege of partnering with one of those executives, Brad Francis, for a couple of years on my journey. Brad understood that achieving consistent positive results in an environment of accelerated change was a non-negotiable requirement for success. He recognized his organization needed to implement tools that embraced change and drove productivity for his corporation to achieve its strategic and tactical objectives. Even more so, Brad believed that to achieve peak results, his people needed to be as important as the project itself and he leveraged every available mechanism to ensure all aspects of his organization were moving in the same direction.

Brad tackles a topic based on a successful career as a seasoned CIO who consistently delivered results based on a pragmatic approach that drove purpose and alignment throughout the organization. To that point, during several meetings at the start of a major project, I recall gathering with Brad to finalize the direction. My organization was the supplier of critical services, and I was engaged directly with Brad to iron out the details of the initiative. We worked our way through resources, timelines, costs, and expectations, among other details, ultimately coming to a unified go-forward plan. What impressed me in those sessions was Brad's drive to make certain everyone was positioned to succeed. That included my organization, his organization, and our collective companies. I later learned of his "do the right thing" perspective toward business and life, and it confirmed what I experienced in working directly with him in those sessions and many others.

The role of a CIO/CTO has become crucial to the success of every company today and aligning the organization to deliver value becomes essential. Constant change creates opportunities, but it also requires us to ensure our teams are on the same page and have the proper mindsets, something that is not an easy challenge for most organizations. Making sure employees are equipped with the right tools has never been more important. *The Alignment Prescription* brings together a simple toolkit that is easily digestible for any technology leader, covering insights (... and hindsight) gathered through years of experience of delivering value and, ultimately, positive results.

Seeing an idea come to life has always energized me. What Brad has provided in this book is an opportunity to keep your employees engaged on the right initiatives at the right time, keeping them synchronized on a unified mission of achievement. No matter what your technology role, you will not go wrong in pursuing the principles of alignment as described herein. This book focuses on maximizing your most valued asset, your people, while driving results through a clearly defined set of principles that scale any organization. It identifies that an accelerated pace of change in today's environment will require technology leaders to implement new tools that drive productivity and ensure their organizations are fully aligned.

Anthony Goonetilleke
Group President – Media, Network, & Technology

"Do the right thing, it will gratify
some people and astonish the rest."

— Mark Twain

INTRODUCTION

In the fast-paced world of technology, creating alignment is critical to the success of any Chief Information Officer and their management team. And alignment is multifaceted, a many-to-many arrangement that must be constantly fine-tuned. Technology organizations must be aligned with the Executive Team, with peer groups and organizations, and with technology employees, staff, and contractors. And with the rapid pace of change impacting the demands on the technology organization, alignment with vendors and suppliers may be similarly important. The pace of change is challenging on everyone up and down the ladder within the technical organization and a lack of alignment will have direct impact on results at the corporate level in the form of profit and loss. Most Chief Information Officers, or Chief Technology Officers, have a seat at the executive table in companies these days, and it is frequently a hot seat. But before we take the deep dive into *The Alignment Prescription*, let me provide a brief background of my journey in understanding its importance.

When I graduated from college and began my first career in teaching, I was naïve about what was necessary to get kids to listen and learn, particularly those in high school. I was teaching mathematics to freshmen, sophomores, and juniors who were all at that stage in their life when they were questioning everything. Why did they need this? Why should they do this? And making things a little difficult for a new teacher was all part of the game they enjoyed playing. To be honest, they did not assign me the advanced placement kids right out of the university student teaching chute, so the challenge really came down to connecting what they were learning to things that they were experiencing in their life, why it was important, and how it could make them better.

As a young teacher fresh out of college in my first real job, the problem for me was that I needed to be doing most of the learning. I did not recognize the importance of connecting the right dots, heck, I did not even know what all the dots were that needed to be connected. And I surely did not know what was required to get there. I had a curriculum, I had subject expertise, I had students, and I had a classroom, but I had no idea how to get them to work effectively together.

As if it were yesterday, I can remember a fellow teacher walking up to me that first day asking if I had my set of rules ready to go, if I had my first week's assignment plan ready, if I had my assigned seating ready, and a list of other "missing" items I should have had ready. The fact is I had extraordinarily little of what was really needed. That afternoon and evening I scrambled around and put together a set of rules, set up my seating charts, and got my assignments ready for the first couple of days. It was a mad scramble, and I did not pull it off all that well. But I can tell you that what I went through those first few days of teaching led to an ongoing and challenging stabilizing effort in year one, a recognition of what planning was going to be required to be successful, and a second year of teaching that was a pleasure. And as for those advanced placement kids, well, I never did get too many assigned to me, but simply because I preferred it that way. And I will leave that as a story for a book on a different topic!

So, what was the difference between year one and year two? The difference was that I learned, because of that first day and through the first year, that from the start I had to be more organized, more prepared, and while considerably basic, first impressions truly mattered. That less than ideal first point of impression caused me to work harder and longer to gain the trust of the students, to demonstrate that I knew what I was doing, and to gain their steady attention. It was much harder than it needed to be in year one and establishing my credibility as their teacher was an uphill battle for a good portion of the year. I was still learning, and they sensed it from the start, and it was difficult to redirect that impression. Ultimately, I figured out how to survive year one of my teaching career, discovered how to not waste time trying to get organized on the fly, and established a level of consistency that ensured my students understood my expectations and what was important for them to succeed in my class.

In year two I entered the classroom with a greater understanding of how to be prepared and how to connect with the students, and it allowed me to focus more on teaching as I eliminated most of the unnecessary disruptions I had previously encountered. Year one I had to focus more on me, but year two was clearly different, a fresh start of sorts. In year two my students and I were on the same page, we had a common understanding of the expectations, and I was able to keep them focused on creating positive outcomes. In essence, this was my first experience with the importance of alignment, understanding the importance of being on the same page with my students, and their parents, and delivering consistent and effective messaging.

Those struggles during my first year helped me enormously through a ten-year teaching career, as a mentor to students, as a coach to student-athletes, and as a partner to school administration and staff. I simply could not expect to achieve optimum results unless I was able to get everyone with whom I was working to be seeking the same goals from the same playbook. And as my teaching and coaching career advanced, I learned that no matter what stage of advancement is the starting point, alignment will move it up a notch or two, almost instantaneously. Getting everyone on the same page increases the success rate dramatically. When I left teaching to begin a new career as a technical professional, mostly due to the desire to increase my earnings potential, those learnings were enormous and had me off and running at a high level. In fact, those same principles of alignment remained with me from my initial technical role as a systems analyst, through many roles of increasing responsibility, ultimately to my roles as CIO and CEO.

Now, before we travel too far down this path of creating effective and efficient alignment, it is important to recognize that you cannot get there without establishing an environment of effective communication and mutual understanding. Alignment is not possible if it is not characterized by consistent messaging that creates an environment of trust. When I left teaching to start a second career, the communication skills I had developed provided a significant advantage in the business world, something I initially overlooked. Likewise, through my past experiences I had learned that it will be very difficult to succeed where trust is not in the equation.

Early in my business career I adopted a "do the right thing" approach. That is not to say I did not make mistakes or veer off track occasionally, but whenever

I struggled with an issue at any level, those four words always grounded me. Many individuals have been cited for that quote, but Mark Twain gets the nod from my perspective. The remainder of the quote was "it will gratify some people and astonish the rest" which, sadly, from one point of view indicates that too many people may not follow the principle. Mark Twain was known as a good person, humorist, author, and speaker. Unfortunately, he was not known for his business prowess, but I will do my best to turn that around for him by leveraging his quote in the spirit of alignment.

One such situation I encountered multiple times in my career were decisions associated with reorganizations, frequently involving layoffs. I recall struggling with the thought of laying off many employees as my company worked its way through some tough times. How do you "do the right thing" and lay off thirty employees? As a business leader, sometimes doing the right thing requires you to take actions that may negatively impact some employees while helping the company achieve its financial goals. Maintaining a positive environment for the rest of the organization will not happen unless there is existing evidence of trust between management and employees. You must be building that trust on an ongoing basis so that there is no question in the minds of others that you are making the right decision. It is a basic alignment principle that, when absent, creates dramatically more angst than necessary.

It is difficult, but you make the right decisions on who will be impacted by whatever criteria is deemed appropriate, often dictated by a Human Resources algorithm. You treat those impacted individuals with respect and provide them with the best available transitional program. It is quite a balancing act to effectively accomplish this under stressful conditions for all involved, but it is much easier in an environment of trust. Trust within your organization is created both by how you function in your role as much as it is by how you communicate. Employees know when other employees are treated fairly, or not. And effectively communicating the decisions and impact to others is equally important. A business mentor, with whom I have great respect as a leader, once counseled me to be "ruthless in your decision making and sensitive in its execution." The bottom line was that I needed to do the right thing, both for the business and those impacted employees, leveraging a focus on effective communication built on a base of established trust. Communication and trust, while often neglected during challenging times, are key ingredients in formulating an effective alignment approach, the focus of this book.

To be clear, this book is written for the busy CIO, CTO, or senior technology leader, who has responsibility for a sizable technical team and an extensive list of priorities and projects to navigate. It is a quick read and provides insight and action steps for those interested in implementing the principles prescribed in the book. While any senior management individual will benefit from the information, I have written this book to provide insight for technology leaders as to the significance of internal alignment in establishing a path toward achieving consistent positive results amidst varied and frequently changing demands. What is different about *The Alignment Prescription* is that it will not focus on the study of business alignment that can be found in a variety of journals or through a typical Google search. It is not hard to find articles identifying the importance of alignment, however, establishing it as a disciplined practice within an organization is a significant challenge.

This book is also not an attempt to explain the concept of business alignment, as most senior technology leaders are already aware of how to define it, even though implementing it is an entirely different challenge. Instead, the focus of this book is on resolving the problem that a lack of alignment creates, working from the inside out, beginning with your own organization. Likewise, the book is not intended to solve the business alignment challenge for your entire company but, rather, to focus you and your technology organization, department, or team on establishing a fundamentally sound approach that is a critical step in creating overall business alignment. It will guide you through a diagnosis of the problem and assist you, as a key technical leader, in developing an alignment approach that meets your specific internal needs. With your own internal alignment challenges cured, the doors will open for broader application across your company, I have witnessed it first-hand. For those interested in implementing *The Alignment Prescription* directly in their work environment, later in the book I provide a software solution for your consideration.

As a point of clarification, *The Alignment Prescription* is being introduced as the first in the

"CIO Rx" series of technology prescription books. Use of the CIO Rx logo, wherever presented, is strictly for the purpose of presenting *The Alignment Prescription* as part of this series. Additionally, if there are any questions about the information provided in the book, feedback you would like to share, or interest in learning more about our software solution to address the alignment challenge, feel free to reach out by messaging me through our website at https://executechsolutions.com/contact.

https://executechsolutions.com/contact

"The main thing is to keep the
Main Thing the main thing."

— Jim Barksdale

CHAPTER 1
THE ALIGNMENT DILEMMA

When you step back and look at the big picture of internal alignment in a technical organization, you can generally classify it in three distinct tiers. The first tier is what we will refer to as the Macro Context, the exterior or outside view of alignment representing a high-level perspective of what is important to ensure everyone has a unified focus. To quote Jim Barksdale, former COO of FedEx, "the main thing is to keep the Main Thing the main thing." Related to *The Alignment Prescription,* that includes an understanding of the projects that are critically important, a recognized prioritization of those initiatives, and the knowledge at any given point in time of the validated level of performance in achieving those projects, as defined. In simpler terms, the Macro Context of alignment represents projects, priorities, and status.

At the other end of the alignment process is what we will reference as the Micro Context, the interior or more intimate tier. The Micro Context represents a specific set of details related to a project which ultimately provides data that is critical for Macro Context reporting. Later in the book we will explore the Micro Context more specifically, but *The Alignment Prescription* relies heavily on an internal understanding of the drivers of performance related to specific projects. Among the Micro Context details would be an understanding of progress in terms of quality, value, and constraints, inclusive of schedule, scope, and cost.

Some technical organizations approach the Micro Context from a predictive perspective. This represents the more traditional approach to project management that drives quality through decision making involving scope, schedule, and cost. Other technical organizations leverage a more adaptive perspective to project management by focusing more on quality and value based on the traditional constraints associated with scope, schedule, and cost. The adaptive perspective most often links to agile project management methodologies as opposed to the more traditional approach. My focus is not in providing an opinion or perspective on one approach versus another, but to provide a mechanism for alignment no matter which approach you have chosen for your organization.

The final tier of *The Alignment Prescription* is what we will refer to as the Meso Context, representing the middle or transitional elements. Well documented requirements, recognition of cross-functional dependencies, and a joint understanding of expectations are key components for organizational alignment at this tier. The Meso Context tends to be where alignment requirements go beyond the boundary of the technology organization and often sets the groundwork for broader business alignment. Its importance should not be overlooked as it encourages engagement with customers, clients, and/or business associates. Despite frequently not recognized as such, the Meso Context creates an opportunity for establishing a level of alignment

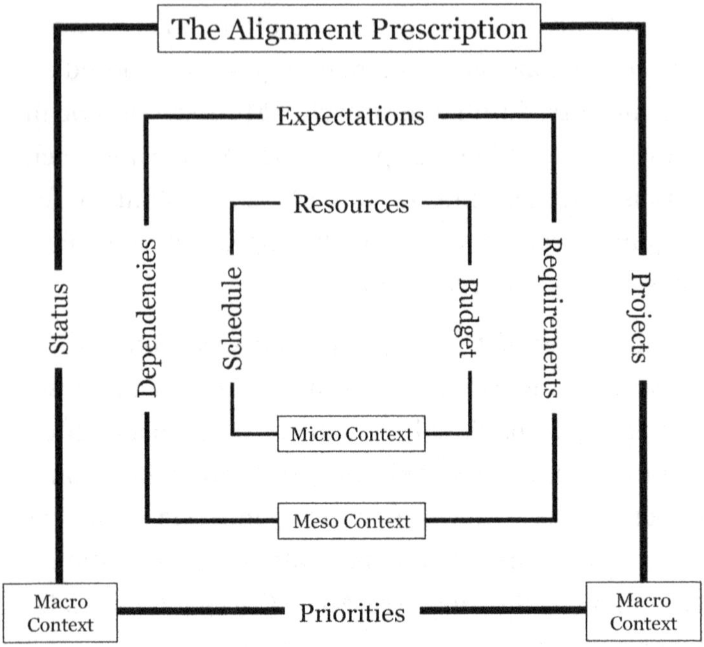

Figure 1

The Alignment Prescription Model

through various business interactions.

A view of the three tiers of *The Alignment Prescription* model, as presented in *Figure 1*, provides a perspective of how each of these elements fit together as the basis of alignment within a technology organization. While each will be explored further in upcoming chapters, the significance of this representation is the knowledge of a requirement to have your hands on the pulse of each of these elements at any point of time. This is one aspect of the alignment dilemma that many organizations experience, the weight of which is frequently underestimated or misunderstood. The default assessment is often a need for improved project management, which in many organizations results in a focus primarily on the Micro Context, at times considering the Meso Context, but rarely addressing the Macro Context.

Your challenge as the CIO, CTO, or senior technology leader in your company is to effectively navigate alignment across all three context tiers. It is to make certain that your organization's priorities are synchronized across all levels and among all contributors. In many regards, alignment is synchronization, a method for getting all contributors similarly informed. Ultimately, the key objective of this book is to provide a solution that places your organization in the best possible position for achieving overall business alignment through a process that focuses on internal alignment first. To that point, resolving the alignment dilemma begins with developing an understanding of the three tier requirements of *The Alignment Prescription* to achieve consistent positive results.

In the upcoming chapters I will frequently reference *The Alignment Prescription* model. For quick access to the visual you may refer to *Figure 1* in this book. Alternatively, if you have not yet taken advantage of the Bonus Offer you may access all the visuals provided in this book, along with other useful information, by downloading *The Alignment Prescription*'s free *CIO Rx Companion Guide* at https://executechsolutions.com/bonus.

https://executechsolutions.com/bonus

"Great performance is one percent vision and ninety-nine percent alignment."

— Jim Collins

CHAPTER 2

THE A5 PRINCIPLE

Through a portion of my technology leadership career, I hung my hat on "People, Process, and Technology" as a mechanism for emphasizing to my technical teams the importance of what the Information Technology organization was responsible for delivering to its business partners. This concept of alignment between information technology and business stakeholders was born in the 1960's but experienced a renaissance in the late 1990's and into the 2000's. I was not alone in leveraging this model as a point of focus in my organization. Over the years, many technical teams in mid-to-large enterprises, various consulting firms, and numerous other technology influencers provided their take on the approach, its meaning, and how it could be leveraged. My organizational pitch at the time was to have our teams strive to be "the premier provider of business solutions through the integration of people, process, and technology." While not unique, it provided an organizational objective that the technical teams were able to easily visualize and pursue.

The common theme, among the vast majority of those who engaged with the People, Process, and Technology (PPT) premise, was to ensure that technology organizations recognized there was much more to their effort than just technology. To that point, people and process played an equally critical role in accomplishing organizational goals. As one way to reinforce the importance during those years, within my organization we recognized individuals and teams for their business and technological success stories in the form of People Awards, Process Awards, and Technology Awards, what we referred to as PPT Awards for short. I have always run with the belief that the key ingredient in every endeavor comes down to people, but even in today's fast-paced world of technology changes, all three components play a significant role and positive outcomes require them to work in unison.

While the drivers of PPT played a contributory role in my management past, I do feel it is important to share an interesting twist to the story. Being excited about rallying my technical teams to focus on the integration of people, process, and technology, I was having a conversation one day with one of my peers who headed up a business unit we directly supported. I was casually explaining the goal of the PPT approach, to be the premier business solutions provider, and he sincerely expressed interest in learning more about it. I later shared some information and slides that I had developed in presenting it to my organization, including a simple Venn diagram (see *Figure 2*) I was using. After discussing the direction and anticipating positive feedback, my business peer looked at me and said, "it looks like you have IT at the center of the universe." We laughed, joked about it a bit, and then moved through the discussion of how I was engaging my team in the PPT program. But I walked away from that conversation with a recognition that as much as

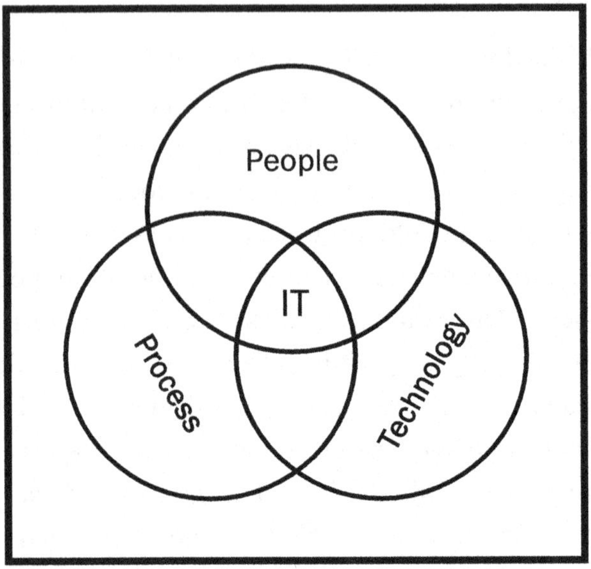

Figure 2

People, Process, Technology (PPT) Model

I was sending a message to my people, I needed to make sure I was taking care of how that message was being received by the business units we were supporting. I needed to work on aligning my internal communication with the message I was sending to the business. To be fair, my business peer was very complimentary of the program we had created and recognized that what we were doing, if successful, would have benefits for his organization. However, while I doubt that he recognized the impact of his comment, it clearly left a lasting impression, and I was determined to not allow it to get in the way of our collective success.

As basic as the PPT pitch may seem, when I look back on it today, it was a significant step toward creating alignment, first from within my organization, but ultimately in conjunction with my business partners. Let us be honest, every CIO, CTO, or senior technology leader recognizes that fully engaged team members will rise to most any occasion and achieve amazing results. Additionally, with the appropriate methodologies leveraged for assessing skills and providing training, people will excel, and your staff will be able to generate a significant transformation to meet the needs of the business. Likewise, no matter what the technology challenge, strong leaders know how to organize the appropriate expertise to address it. But where the bumps on the road tend to surface is when processes do not effectively provide for people and technology to find the most effective and efficient path to the finish line. This is where the need for optimizing or reengineering existing processes may need to be a consideration. That said, often the capability for improving processes and creating a more effective and efficient program are within reach if you look in the right places.

In the subsequent chapters of this book our focus will be on processes primarily impacting internal alignment, taking into consideration the three tiers of context that we previously identified: Micro, Meso, and Macro. But it needs to be clearly recognized that by focusing on internal alignment, I am highlighting the importance of being on the same page from top to bottom in your organization to achieve consistent positive outcomes. This is an important prerequisite to the goal of ultimately being aligned across the entire business enterprise. In the words of Jim Collins, author of *Good to Great*, "great performance is one percent vision and ninety-nine percent alignment."

The CIO, together with the senior staff, must translate the strategic objectives of the enterprise into recognizable technical goals and related initiatives. What the CIO views as important initiatives must be clearly understood by each level of management and their teams. For the technical organization, this really begins with establishing a common understanding of the Macro Context of projects, priorities, and status. The CIO must establish a mechanism of confirmation as to what everyone recognizes as their top projects. They must ensure that their technical teams understand how these projects contribute to organizational goals. And, finally, the CIO must ensure the organization is consistently progressing toward achieving positive results aligned with these goals.

The A5 Principle (A5P), as exhibited in *Figure 3*, is the idea that to achieve the strategic objectives of the enterprise, as translated to a set of important technology initiatives, the CIO must leverage a consistent management approach that creates awareness, ensures accountability, drives achievement, and solidifies internal alignment. Absent of one or more components, A5P concludes that technical organizations will struggle to achieve

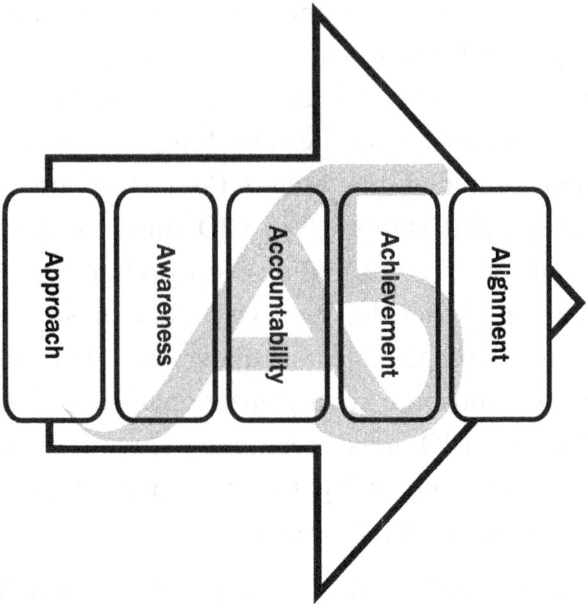

Figure 3

The A5 Principle

broader business alignment. In their book *Alignment*, Robert S. Kaplan and David P. Norton write that "enterprises must have active policies to communicate, educate, motivate, and align employees with the strategy. They must also align their ongoing management processes – for resource allocation, target setting, initiative management, reporting, and reviews – with the strategy."

The A5 Principle specifically addresses this identified need, that is, the requirement for an ongoing management process within your technology organization. For example, by applying the components of A5P to the Macro Context - projects, priorities, and status - and establishing a mechanism to make that information universally accessible, you open a window of knowledge for individual contributors, teams, and management. It plays a significant role in keeping technical teams unified in driving toward consistent positive outcomes that are synchronized with organizational goals. Simply put, you create a level of internal alignment. While one example of many that we will explore, this is the critical process ingredient in *The Alignment Prescription*.

So, to understand A5P in more detail, let us take a

closer look at the components as applied to the Macro Context representing your organizations' projects, priorities, and status. At the front-end of A5P is the need for an approach that provides clarity around what projects are more important than others and how they are progressing. The approach, on one end, creates the direct line through to the alignment we seek at the other end of A5P. While some form of roll-up status reporting or online project information may provide the CIO with a sense of what is being accomplished within their organization, it frequently does not provide insight into the priority conflicts that stymie progress.

In an ideal sense, the approach should capture the prioritization of the projects at all levels of the organization. It should identify the priorities of the individual, management, and the senior technology leaders, while being visible to as broad a set of contributors and interested parties as is feasible. The more visibility into the priorities as identified by each contributor, the more likely that priority conflicts are quickly identified, and priority conflicts resolved. In contrast, the longer that priority conflicts sit unresolved then the more likely internal alignment is disrupted. These conflicts often surface in employee surveys or, more damaging, at the water cooler with the "my priorities are always changing" dialogue. They often show up in project outcomes in the form of delays, financial challenges, or various other disappointments. Establishing an approach that creates transparency regarding each contributor's priorities eliminates a significant underlying barrier to positive outcomes, an aspect which is often overlooked.

The approach becomes even more valuable if it can extend its transparency on priorities to also include more details about the projects and their status. Imagine if everyone in your organization was able to simultaneously view each contributor's prioritized projects and understand the status of their progress toward established goals. With a well-designed approach this is indeed achievable, and it begins to generate the awareness, accountability, and achievement that is required for collective success. The APPROACH creates the AWARENESS, the awareness ensures ACCOUNTABILITY, and the accountability drives ACHIEVEMENT. This is indeed the direct line that is created by establishing an effective approach to achieve the desired ALIGNMENT. It is the A5 Principle in action, and it will increase your organization's ability to achieve consistent positive outcomes. It is important to recognize that thus far we have only focused on the Macro Context as it relates to the A5 Principle, a segment of the overall remedy provided in *The Alignment Prescription*. In the ensuing chapters we will apply the concepts of the A5 Principle to the Meso and Micro Contexts as well.

"Visualize your goals and ask yourself What's Important Now. Think WIN."

— Lou Holtz

CHAPTER 3
THE MACRO FOCUS

By now you have recognized that *The Alignment Prescription* places a significant amount of emphasis on priorities. Internal alignment has a lot to do with establishing effective IT governance, of which prioritization plays a key role. But too often at the enterprise's Macro Context level, project management plays a significant role and exhorts a sizable investment, while management of priorities is frequently viewed as incidental. Priority management is often dismissed as automatic, frequently assumed to require a minuscule investment of time and resources, and simply a process of passed-down communication. Now, this is not to say that the investment of time and resources for project management and priority management need to be equal or reversed, but their position of importance within the enterprise deserves cautious reflection.

In the late 90's, I was fortunate to be in the audience for a motivational speech conducted by the legendary author, football coach and sports commentator, Lou Holtz. This was an annual event for top performers within my company, internally billed as the President's Achievement Award. The recognition event during this particular year took place in Hawaii and, among the activities that were required participation events, this one was directed at leadership. Coach Holtz was the guest speaker and, if you have never experienced the entertaining philosophical insights of Lou Holtz, you should check out one of his videos from similar leadership events that are readily accessible online.

That evening, Lou spoke on leadership philosophies for about an hour, a portion of which he discussed his perspective regarding "What's Important Now?" or, as he says, WIN. The brief synopsis of this approach was to challenge his players to ask themselves that specific question multiple times per day and answer it by making winning decisions. Coach Holtz would have his players set their goals and, each time they had something to decide, ask themselves "What's Important Now?" It did not matter if it was as simple as, "Should I eat breakfast or not," or as tempting as, "Should I party or study," his challenge to his players was to first ask themselves that question with a winning thought process in mind. The idea was to keep them fixated on the things they needed to accomplish to be successful in football, their personal life, or anything for which they set specific goals to achieve. The importance of this WIN philosophy for his players was to create focus, build a winning attitude, and get better every day.

That speech by Coach Holtz left a significant impression on me. It was a mindset I felt could translate to my role as a technology leader to improve the overall effectiveness of my organization. Instead of the team on the playing field, it made sense to me that the WIN philosophy could apply to my technology team, albeit with some adaptation. Of course the components of the business decision process would be different, but the thought process in determining what was important was still about establishing a winning structure for the organization and each of the individuals involved.

Initially I introduced Coach Holtz's WIN philosophy in team meetings as a form of self-motivation and focus. It was well-received from an individual perspective, but it was not the business driver I envisioned. My thoughts were to incorporate it more broadly to drive team results, much like I envisioned Coach Holtz would have done in building winning football programs. To that end, over time I decided to link "What's Important Now" not only to individual decision making but more broadly across our technical teams as a philosophy of prioritization. Two factors influenced this thought process, one at the individual contributor level and one at the senior leadership level.

At the individual contributor level, most technology leaders have experienced the "they keep changing my priorities" or "what's today's number one" condition expressed by an employee. Perhaps you are working with a team member who is struggling to keep up with their workload. Examples of this may be situations involving under-resourcing, a skills deficit, or workload management. These conditions frequently relate to contributors who struggle to complete initiatives within expected timeframes, budgets, or quality levels. Hence, they may default to expressing concern for changing priorities influencing their ability to get things done as expected. It is difficult to get a clear view into the experience that a contributor is describing without shadowing them in their role, which simply is not practical in most cases. Hold that thought and we will come back to providing a solution to this shortly.

The second factor influencing this WIN philosophy of prioritization was at the senior leadership level. As CIO, it was clear to me what my organization's priorities needed to be to address the needs of the key stakeholders. I made it a point to be clear about those priorities as I communicated them to my senior leadership team. Where the vision of prioritization began to get less clear was related to how my communication was translated to all areas of the technical organization. Strategic goals were well publicized, communication was structured and consistent, and incentives were connected to organization goals. It seemed like a well-tuned engine, albeit that it was being observed while standing outside the car with the hood closed. The key question was whether the CIO priorities were effectively translated to each member of the technical organization. To continue with the car analogy, was everyone who was responsible for some aspect of the vehicle focused on moving it toward the same destination? Were everyone's priorities synchronized?

As described, both the individual contributor and senior leadership factors associated with prioritization presented an opportunity for a common resolution. Now your priority alignment conditions may not be identical to those described, but the remedy is universally applicable. To ensure internal alignment, a consistent understanding of priorities among all levels of the organization is necessary. Whether you are the software developer, network engineer, or a member of the management team, having visibility and understanding of how each contributor views what is important is critical to success. Consistent with the A5 Principle, you must establish an approach that creates awareness, ensures accountability, drives achievement, and solidifies internal alignment. Having a high degree of priority transparency at all levels will enable the organization to continue to move forward in a cohesive manner, making even the most difficult challenges achievable. The WIN philosophy works when everyone is working on the most critical initiatives which, as an ingredient in *The Alignment Prescription*, is bolstered through prioritization transparency.

Now, in the business world there is a broad range of analysis that goes into determining the most important initiatives at any given point in time, as depicted in *Figure 4*. As the CIO, CTO, or senior technology leader, you evaluate a range of inputs including strategic perspectives, executive directives, client needs, budget availability, required timeframes, resource capacity, technology constraints, risk management, and available skillsets. Depending on where a contributor sits in the technical realm, the approach to prioritizing is influenced by factors that may not be evaluated equally by other constituents. To this point, it is no surprise that the CIO will likely view priorities through a different lens than the software developer. This is a natural occurrence and by no means intended to represent a problem. Everyone will view what is important based on factors associated with their role and responsibilities. Understanding how

Figure 4

What's Important Now (WIN)

everyone identifies their priorities, as portrayed in *Figure 5*, will drive opportunities to align among those varied roles. Said differently, having visibility into how each contributor views their priorities provides an opportunity for recommending adjustments. Priority visibility and alignment are critical in keeping an organization focused on achieving positive outcomes, both in an immediate sense, as well as to longer-term goals.

As presented, one of the key internal alignment challenges is to synchronize priorities across all areas of the technical organization and to create the visibility, or transparency, that enables everyone to understand how others visualize what is important. To that point, each individual contributor needs the feedback necessary to ensure they are working on the right things at the right time. All of this has become even more challenging with the growth of remote work arrangements, most typically involving employees working from their home. Less natural interaction and direct visibility into progress being made by a contributor's manager, peer, subordinate, or a related project team, makes it more difficult to keep everyone on the same page. But whether they are remote or in the next cubicle, keeping all

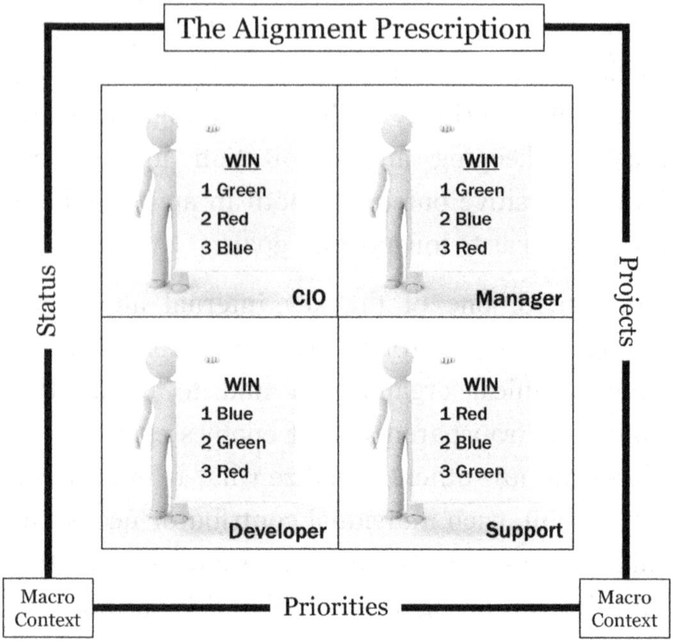

Figure 5

Priority Alignment Challenge

employees engaged and synchronized with organizational objectives requires more than any of the typical communication methodologies provide.

In many organizations this is accomplished through varied communication mechanisms such as management information sessions, periodic project reviews, employee performance evaluations, and status reporting, to name a few. Other organizations provide input on strategic initiatives that management translates into a set of required projects to achieve organizational goals. Still others assign project management positions with responsibility for maintaining the organization's list of critical projects. These positions typically meet with key contributors on a scheduled basis and provide valuable feedback on progress to technology management, as well as senior technology leaders. Off course, there are many other variants to the management of top priorities, too many to list here. No matter what approach is leveraged, your critical internal alignment objective must be to ensure that the priorities of the organization are recognized, that individual contributors have visibility into the related projects, and that the status and progress of these initiatives are understood by all stakeholders.

Projects, priorities, and status are what defines the Macro Context of *The Alignment Prescription*. As described, by establishing a high level of transparency through a system of priority management, the CIO, CTO, or senior technology leader takes an important and urgent step in aligning their organization internally, creating a baseline for broader business alignment. To that end, the priorities, projects, and status of the Macro Context relies upon critical data provided from both the Meso and Micro Context levels. In the next chapter we will begin to explore this relationship beginning with the requirements, dependencies, and expectations of the Meso Context.

“People don’t do what you expect
but what you inspect.”

—Louis Gerstner

CHAPTER 4
THE MESO LINKS

Most of you will be familiar with a well-leveraged quote regarding expectations: "People don't do what you expect but what you inspect." There are many variations of the statement and plenty of debate as to where it originated. Most credit goes to Louis V. Gerstner, Jr. from his 2002 book about IBM's historic turnaround called *Who Says Elephants Can't Dance*. Gerstner offers great insight into management, leadership, and culture, a recommended read, but for the purpose of this book we will simply leverage a short and sweet variant of his quote, "expect what you inspect".

In discussing the Macro Context of *The Alignment Prescription*, we explored the impact of priorities. As we move to the Meso Context, the middle or transitional tier, we take what has been identified as important at the top level and connect it with their associated requirements and dependencies. It is here where we place more definition to the expectations by making certain there is complete understanding of what is required to achieve them. Without clear expectations of how everyone contributes to the success of an initiative, the initiative will get bogged down with, at a minimum, endless communication snafus. In contrast, investing time up front by clarifying requirements and understanding dependencies establishes the baseline from which effective communication will commence.

The Alignment Prescription recommends that there be an increased focus on cross-functional dependencies, those not within direct control of the project owner. Typically, individual contributors to an initiative will have a well-organized understanding of what they are responsible for completing and who within their internal team must deliver them. The key to success is to identify the external inputs that are required, those for which there is not direct control, and ensure there is a common understanding of the deliverables. To the project manager, and others with an ownership stake, these are the cross-functional dependencies for which they must rely upon others to keep the initiative moving in the right direction. These cross-functional dependencies represent a potential risk to achieving success. Because it is not within the direct control of those with an ownership stake on a project, there is a tendency for it to become a convenient justification for, as examples, scheduling or budget misses. The expectation must be set that this needs to be monitored effectively, as is the case for those items under their direct control. The best way to eliminate the potential for an embedded excuse is to close the opportunity for allowing it. We will discuss more regarding the best way to manage this but suffice it to say that fully understanding dependencies and requirements are critical to achieving success.

To be clear, the dependencies we are discussing as part of an internal alignment process are primarily resource based, not task dependencies created in a project plan. In this sense, alignment is a person-to-person challenge, not a task-to-task one. That said, it is important that the task owners are clearly identified so that all participants recognize their contributory role. The critical nature of identifying the stakeholders and contributors for the prioritized initiative cannot be emphasized enough, including those outside the directly managed scope. To achieve a positive outcome you must ensure there is buy-in among all parties related to their responsibility for specific deliverables.

Again, be certain to perform a complete rundown of all contributors and inventory them so there is no doubt of the scope of involvement of everyone, particularly those who cross over into other disciplines. The goal is to ensure these cross-functional dependencies do not become a burden to the core team responsible for the success of the project, while at the same time holding them accountable for identifying and tracking them. Under most conditions, a critical cross-functional dependency will be the definition of requirements for a specific project. Project teams typically rely on contributors beyond their own organizational boundary, the requesters, to provide the objective and definition of what needs to be accomplished. Let us be honest, projects are often initiated without a full and documented understanding of the requirements, so it is critically important that the responsibility for providing them is understood by all involved parties. The bottom line is that dependencies and requirements are the prerequisite of establishing clear expectations on what is to be delivered and who is responsible for delivering them.

The inspection challenge related to expectations is that as the CIO, CTO, or senior technology leader, you have plenty on your own plate without having to take on the burden of constant monitoring. It is unlikely that you have the time or capability to stay focused on the inspection of ten, twenty, thirty, or more initiatives simultaneously. So how do you inspect without obsessing over it? Simply stated, you put the control into the hands of those reporting it by establishing a process of self-monitoring. But self-monitoring alone is not sufficient to get the result you expect, thus requiring you to inspect by adding the element of visibility. You identify what is important and you empower others to pull together the requirements, identify the dependencies, and you make their progress visible and transparent. We will address how to create this visibility and transparency in later chapters but before we move on, let us look at another important element about the relationship

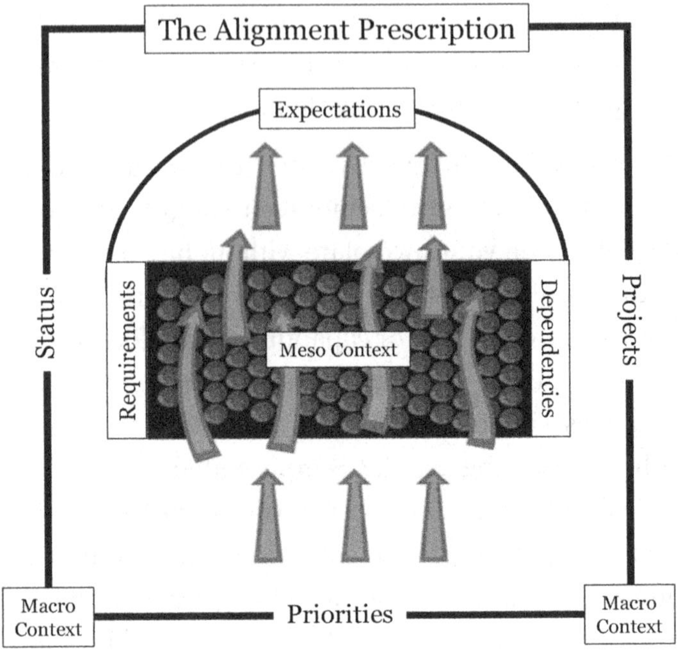

Figure 6

The Expectations Filter

between priorities and expectations.

In applying *The Alignment Prescription*, you make your priorities visible and, by doing so, you create an expectation of delivery among stakeholders that must be addressed. It is for this reason that the need to create shared accountability through the detailing of requirements and acknowledgment of dependencies becomes critical. *Figure 6*, the Expectations Filter, depicts how the priorities you have defined lead to the expectations that are not always in your direct control. But by filtering your prioritized projects through a process of detailing requirements and identifying core dependencies you normalize expectations cross-functionally. The priority is no longer only your priority, it is the priority of those that extend beyond your structural or management boundary. In essence, this creates a natural alignment and shared accountability across your organization. Likewise, to the degree there is visibility beyond your own organization it creates broader alignment.

Another aspect of this filtering process is that of communication. Things do change and often at a rate of speed that may cause contributors to get sloppy and cut corners, particularly as it relates to communication. The goal should be to create points of interaction that are natural, repetitive, and seamless. Communication is something that we all do every day in our personal and professional lives. We take it for granted until something goes dramatically wrong. Then we try to straighten out the problem, which some will refer to as a breakdown in communication, by communicating some more. So it is important to provide mechanisms that avoid the potential of any breakdown.

In the informal sense, on-going communication represents the day-to-day verbal interaction we practice, often unplanned, in the name of getting our job done. In the formal sense, on-going communication represents scheduled updates related to a multitude of areas including, but not limited to, unforeseen changes, expectations versus performance, and progress on initiatives. Within the realm of expectations, this includes the need to communicate changes in dependencies or requirements to ensure all stakeholders have an equal understanding of progress and changes.

For ongoing alignment, it is best if communication surfaces automatically so that it is practiced without special circumstances requiring it. You need a communication process that ensures visibility of what is important and exposes any barriers in achieving it. This is the unique communication process that is an important ingredient in *The Alignment Prescription.* It represents a merging of the informal and formal processes and differentiates itself from more reactive approaches. To aid with this process I will introduce a toolset for your consideration in a later chapter, however it should be recognized that this communication process can be as simple as establishing a periodic scorecard or status report. Such a scorecard needs to provide stakeholders with consistent updates on the priority and status of a project while being cross-functionally focused on any requirement or dependency gaps. At a glance, this type of scorecard must provide a solid perspective of the overall state of the project. *Figure 7* is provided as an example of one of many representations of a possible scorecard that includes the key elements of the Macro and Meso Contexts.

Alignment requires a consistent information flow, so it is critical to determine how to communicate effectively to ensure that expectations are clearly understood by all involved parties. The bottom line is that people do not naturally want to report on

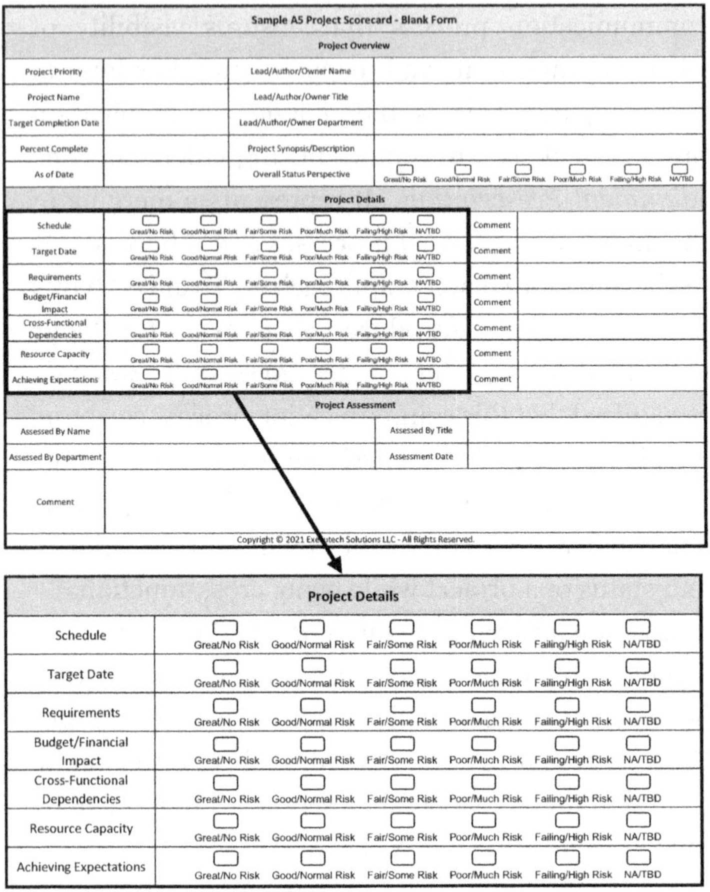

Sample A5 Project Scorecard - Blank Form

Project Overview

Project Priority		Lead/Author/Owner Name	
Project Name		Lead/Author/Owner Title	
Target Completion Date		Lead/Author/Owner Department	
Percent Complete		Project Synopsis/Description	
As of Date		Overall Status Perspective	☐ Great/No Risk ☐ Good/Normal Risk ☐ Fair/Some Risk ☐ Poor/Much Risk ☐ Failing/High Risk ☐ NA/TBD

Project Details

Schedule	☐ Great/No Risk	☐ Good/Normal Risk	☐ Fair/Some Risk	☐ Poor/Much Risk	☐ Failing/High Risk	☐ NA/TBD	Comment	
Target Date	☐ Great/No Risk	☐ Good/Normal Risk	☐ Fair/Some Risk	☐ Poor/Much Risk	☐ Failing/High Risk	☐ NA/TBD	Comment	
Requirements	☐ Great/No Risk	☐ Good/Normal Risk	☐ Fair/Some Risk	☐ Poor/Much Risk	☐ Failing/High Risk	☐ NA/TBD	Comment	
Budget/Financial Impact	☐ Great/No Risk	☐ Good/Normal Risk	☐ Fair/Some Risk	☐ Poor/Much Risk	☐ Failing/High Risk	☐ NA/TBD	Comment	
Cross-Functional Dependencies	☐ Great/No Risk	☐ Good/Normal Risk	☐ Fair/Some Risk	☐ Poor/Much Risk	☐ Failing/High Risk	☐ NA/TBD	Comment	
Resource Capacity	☐ Great/No Risk	☐ Good/Normal Risk	☐ Fair/Some Risk	☐ Poor/Much Risk	☐ Failing/High Risk	☐ NA/TBD	Comment	
Achieving Expectations	☐ Great/No Risk	☐ Good/Normal Risk	☐ Fair/Some Risk	☐ Poor/Much Risk	☐ Failing/High Risk	☐ NA/TBD	Comment	

Project Assessment

Assessed By Name		Assessed By Title	
Assessed By Department		Assessment Date	
Comment			

Project Details

Schedule	☐ Great/No Risk	☐ Good/Normal Risk	☐ Fair/Some Risk	☐ Poor/Much Risk	☐ Failing/High Risk	☐ NA/TBD
Target Date	☐ Great/No Risk	☐ Good/Normal Risk	☐ Fair/Some Risk	☐ Poor/Much Risk	☐ Failing/High Risk	☐ NA/TBD
Requirements	☐ Great/No Risk	☐ Good/Normal Risk	☐ Fair/Some Risk	☐ Poor/Much Risk	☐ Failing/High Risk	☐ NA/TBD
Budget/Financial Impact	☐ Great/No Risk	☐ Good/Normal Risk	☐ Fair/Some Risk	☐ Poor/Much Risk	☐ Failing/High Risk	☐ NA/TBD
Cross-Functional Dependencies	☐ Great/No Risk	☐ Good/Normal Risk	☐ Fair/Some Risk	☐ Poor/Much Risk	☐ Failing/High Risk	☐ NA/TBD
Resource Capacity	☐ Great/No Risk	☐ Good/Normal Risk	☐ Fair/Some Risk	☐ Poor/Much Risk	☐ Failing/High Risk	☐ NA/TBD
Achieving Expectations	☐ Great/No Risk	☐ Good/Normal Risk	☐ Fair/Some Risk	☐ Poor/Much Risk	☐ Failing/High Risk	☐ NA/TBD

Figure 7

Sample A5 Project Scorecard

things going astray, so creating this type of ongoing process negates the temptation of holding back less than positive results. Understandably, I recognize that there are hundreds of scorecards and status reporting variations that exist, and this need is neither new nor exciting. But we are looking at alignment as a twist on the normal scorecard objective and we need to ensure that its deployment is simple and self-managing. So, resist the temptation to overengineer that need since you merely want to ensure expectations are aligned through a validation of the requirements and dependencies. In simpler terms, you just want to inspect what you expect. Now having explored some of the most overlooked elements of achieving alignment as exhibited within the Macro and Meso Contexts, the next chapter will take us through the remaining ingredients of *The Alignment Prescription* as we explore the Micro Context.

Before we move on, it is important to note that the sample scorecard is not intended to represent a solution, but rather exhibits the key elements that need to be considered within any alignment-oriented scorecard process. Thus far we have only addressed a portion of the elements that you see exhibited on the A5 Scorecard, however the sample does provide an appropriate visual of how all three contexts work together as part of *The Alignment Prescription*. If you are interested in obtaining a copy of the Sample A5 Project Scorecard as shown in *Figure 7*, both a blank form and an example of a completed scorecard are accessible for downloading by using the following URL: https://executechsolutions.com/download.

https://executechsolutions.com/download

“Less is more only when
more is too much.”

—Frank Lloyd Wright

CHAPTER 5

THE MICRO LIMITS

Embedded in the thought process for introducing a project is the assessment of the impact to the resources, schedule, and budget for which it will be assigned. In the traditional sense, given clear requirements, it is reasonable to expect proficient technical professionals to estimate the cost and timeframe for a project based on established quality assumptions. Add more to the scope of the project and it will generally impact the time required, the cost of the project, or both. In the Meso Context we placed emphasis on requirements, dependencies, and

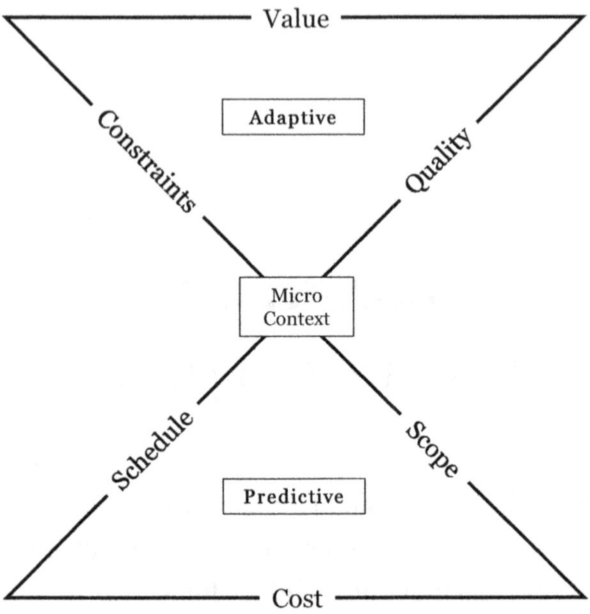

Figure 8

Micro Context Elements

expectations for good reason. Changing requirements or dependencies have the capability to create a variable influence on the resources, schedule, and budget of a project. These Micro Context elements are critical to monitor in a synchronous manner to ensure one does not inappropriately influence the others.

As project management approaches have matured over the years, including the evolution of agile methodologies, the emphasis on the Micro Context has shifted to some degree, but it has not eliminated the importance of dedicated resources, an established budget, and a predictable timeline for delivery. Where scope, schedule, and cost have traditionally been the most important factors in managing the quality of outcomes, many organizations have considered a more adaptive approach involving an ongoing assessment of quality, value, and constraints. With this paradigm shift, in many development efforts instead of the scope driving the time and cost, time and cost drives the scope. Many consider this to be a value driven approach leveraging a fixed cost and fixed timeframe to create feature estimates, turning the traditional approach upside down, so to speak. *Figure 8* illustrates key differences in the perspective of adaptive versus predictive methodologies, but it is important to understand that the Micro Context elements of resources, schedule, and budget are embedded independent of the approach, often under the umbrella of quality. Let us take a closer look at this.

With fixed costs, fixed timelines, and a variable scope, risks decline over the course of the development effort. On the flip side, it is less clear as to what will be delivered on the due date although, again, that risk is reduced as the project evolves. Where some concrete deliverable is not the primary obligation; the acceptable variability of the adaptive approach may better manage the constraints of cost and schedule. Arguably, where scope is identifiable and specific, the predictive approach may best manage within established cost and time constraints. But under all conditions, schedule and budget still play a vital role, particularly from the perspective of key stakeholders.

Depending on the circumstances of your product, services, and delivery philosophy, there is a place for any one of the wide varieties of development and project management approaches. While in today's quick changing environment many technical organizations see advantages in moving to more agile approaches, there are just as many organizations effectively leveraging waterfall methodologies or some form of hybrid variation. When it comes to business alignment, organizations should not feel boxed in with one software development or project management approach over the other. The bottom line is that the chosen approach is not the driver of business alignment that many believe it represents.

The CIO, CTO, or senior technology leader striving for alignment must recognize that its achievement is multifaceted. Many organizations over the past decade have moved to implement agile teams only to discover that, while it had an influence on how projects and deliverables are defined and delivered, it did not create greater business alignment. Typically development teams are well aligned within their specific structure, but overall internal and business alignment is a much greater challenge. This is not to say that there is not value in striving for a specific development structure, it is simply a statement that the approach alone will not address the alignment dilemma we seek to resolve.

No matter what software development and project management approach your organization chooses to pursue, waterfall through agile, the ingredients for

The Alignment Prescription are essentially unchanged. Certain elements of the Micro Context may be weighted as more significant than others, but there is a consistency in monitoring them that must be achieved. No matter what methodology you choose to have your organization leverage, you will not be surprised to discover that there will be many interested parties who will be looking for a timeframe for delivery. And if there is any thought that you will get a pass on budget, clearly that will not fly far with those interested in the financial management of an initiative. The point here is that alignment is broader than the methodology you choose for software development or project management, and synchronization around the Micro Context represents an important element of visibility among all stakeholders. When you provide insight into the resources, finances, and timeline of an initiative to interested parties, you create another level of alignment.

So how do we create alignment around schedule, budget, and resources within the Micro Context? I have frequently emphasized communication and transparency as key components of *The Alignment Prescription*. Under an assumption that there has been adequate planning, we have addressed the need to monitor requirements and dependencies within the Meso Context. Similarly you must build upon these areas to provide enough detail to keep everyone informed of progress, while avoiding getting trapped in enormous detail. One of the senior executives in a company for which I worked would frequently state that "less is more". While the root of that statement is in architecture, although now leveraged for a variety of purposes, his usage primarily pertained to communication up the organizational ladder under a belief that more information opened the door to more questions. While I never fully adopted that principle, I did keep it in the back of my mind. The philosophy was applicable in some executive interactions, but there clearly were others that simply wanted more information and to them, less needs more. In the words of Frank Lloyd Wright, "Less is more only when more is too much."

While those differing perspectives tend to contradict each other, collectively they make the point of communication and transparency as it pertains to alignment. Somewhere in the middle is most often the appropriate place to land, so describing your progress in terms of the Micro Context elements as shown within the A5 Scorecard (previously presented in *Figure 7*) is the point of balance we seek to achieve in *The Alignment Prescription*. You want to keep your stakeholders aware of schedule, resource, and budget in an easy-to-follow manner. You therefore provide information at a higher level in terms of accomplishment and risk (less is more), but you provide context through comments that satisfy the need to understand any changes (less needs more). Most importantly, you provide information as to any variances as soon as they are recognized, thus building trust around your communication and transparency. If nothing has changed between updates, then the need for comments is limited, if at all. But if something has created a status shift, then a brief comment keeps everyone aligned.

Those desiring more information will initiate that discussion and those satisfied with the information received will feel informed. Finding the right balance of information to communicate to key stakeholders is another key ingredient of *The Alignment Prescription*, although admittedly a bit more difficult to define with precision. Keep in mind that your comments represent a historical view of the progress of an initiative, so make them clear and precise, avoiding any form of tainted truths that may get in the way of creating the alignment you seek.

We have journeyed the distance related to the contexts of *The Alignment Prescription*. At the inner core is the Micro Context representing the schedule, budget, and resources. It is influenced by information provided from the Meso Context, specifically requirements and dependencies, and it generates information to the Meso Context that impacts expectations. Those contexts are surrounded by the overall influence of the Macro Context that generates a collective set of projects, priorities, and their representative status. Collectively, these contexts create the structure of *The Alignment Prescription* based on the fundamentals of the A5 Principle, specifically the approach that generates awareness, accountability, and achievement leading to overall organizational alignment.

In the next chapter I will address one additional ingredient influencing alignment, organizational structure. While the structure of your technology organization is not displayed as part of *The Alignment Prescription* Model, it does significantly influence how you interact internally and with business partners.

"An organization, no matter how well designed, is only as good as the people who live and work in it."

— Dee Ward Hock

CHAPTER 6

THE ALIGNMENT STRUCTURE

There are many ways in which technology organizations are structured from the basic hierarchical or functional approach to what is sometimes referred to as an open organization. Likewise, there is plenty of justification that can be generated supporting one structure versus another. It is not the focus of this book to attempt to justify any single approach to building your organizational structure, but it is important to know what variables you will need to address to ensure effective

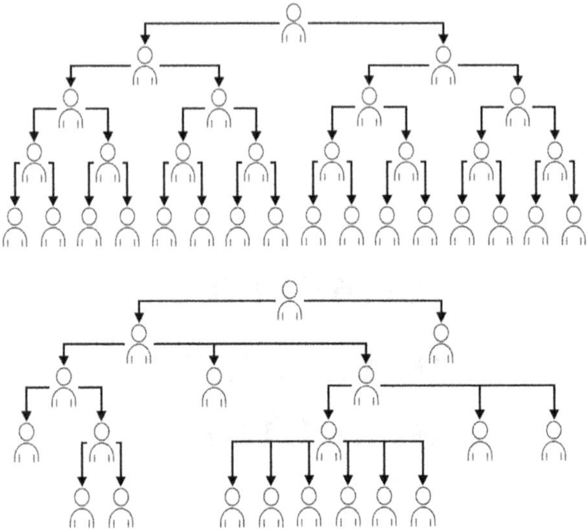

Figure 9

Functional Organizational Structure Examples (Hierarchical)

alignment. Depending on the author, you may discover perspectives identifying three, four, five, or more unique models that apply to technical organizations. I am going to limit the list by focusing on several of the more commonly leveraged structures and explore the relationships that directly impact internal alignment.

To begin, let us look at the one-to-one relationship between an individual contributor and their boss in what is described as the functional or hierarchical organizational structure as shown in *Figure 9*. Each individual reports directly to a single leader, manager, or supervisor. Generally, at the top of the structure is an individual such as a CEO, President, or Chairman. Different parts of the organization are most often departmentalized such as with marketing, finance, operations, etc. The beauty of this structure is that there is great efficiency in decision-making and direct lines of communication are highly visible and well understood. On the challenge side of the design is that it creates some natural barriers for cross-functional communication and there is a tendency for decisions to fall within the hands of a few. For the individual contributor, the business relationships are clearly defined with almost all direction being provided by one direct manager. To that point, it is easy for an employee to understand their position within the overall structure of the business, as well as their typical points of interaction with other members of the organization.

From an alignment perspective, an increased focus on the Meso Context is critical in addressing the departmentalized impacts of the hierarchical structure. Cross-functional dependencies must be clearly understood and communicated, requirements must be unambiguously defined, and project teams must proactively seek input on expectations. Among the reasons that organizations tend to depart from the hierarchical structures is due to its inherent limitation on collaboration, which can be minimized by focusing attention on ensuring the needs of the Meso Context are emphatically pursued. With decision making in the functional structure representing a demonstrable strength, many organizations recognize the critical nature of implementing strong supplemental cross-functional communication processes to ensure the challenges of the structure are minimized.

In some companies, striving to be more horizontal in structure leads to what is frequently referred to as a flatter organization. Essentially, the flat organization endeavors for fewer levels and encourages open communication. There are many different types of flat organizational structures leveraged by companies, some more flat than others, and the relationship between an individual and their boss, or bosses, varies greatly. *Figure 10* exhibits a few conceptual variations of the horizontal or flat structure. It would be common for an employee to have a one-to-many relationship with management, typically seeking input based on the expertise of leadership instead of being dependent on a single direct management relationship. Some flatter structures strive to create a more projectized design, where the project drives its creation as compared to the more functional approach.

No matter how the horizontal structure may be defined, it should be noted that it is not void of hierarchy, it is simply minimized in the spirit of employee engagement, team development, and collaboration. This minimization is frequently fostered by the reduction or elimination of the middle-management layer. Many companies look at this structure as a balance between hierarchical and other more revolutionary organizational concepts,

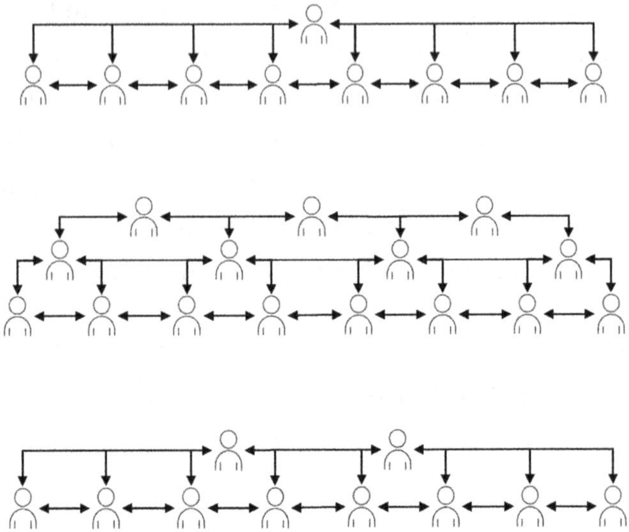

Figure 10

Flat Organizational Structure Examples (Horizontal)

providing them the opportunity to empower employees while still retaining aspects of traditional management to the degree they deem effective.

While there are many benefits to flatter structures, given that it creates less direct management influence, it risks some loss of control overall. As an example, when initiatives go astray there is a greater potential for the condition to not rise to the attention of management in an expedient manner, placing an increased need for attention at the Micro Context level. With more individuals involved in project assessments it becomes crucial to avoid a paralysis of decision-making. The structure creates a natural tendency for broader employee involvement with less management oversight, so keeping a sharp eye on variabilities to schedules and budgets is important. These challenges are offset by having fewer managers involved in the decision-making process which reduces the micro-managing condition that many hierarchical structures tend to create.

Weighing strengths and weaknesses, this is the direction that a significant number of companies are establishing as their operating norm. This tends to be an appealing structure for a CIO, CTO, or senior technology leader who strives for increased collaboration among their teams without radical changes to a functional organizational design. That said, from the perspective of alignment, the horizontal structure does retain the same Meso Context challenges of the hierarchical structure. Outside the boundaries of the teams there remains a substantial need to manage expectations and ensure cross-functional dependencies are understood.

A point of caution related to flatter structures is often the reason for its interest. Frequently, flatter structures are pursued in the spirit of cost reduction or containment, eliminating a layer of management, and/or expanding the number of direct reports for each member of the management team. From the alignment perspective, this frequently dilutes the points of interaction by placing added demand for communication on a smaller number of management contributors. Regardless of the increased attention given to a specific context level, if your move to a flatter structure is implemented as a wider hierarchical structure, then alignment will suffer.

Beyond the frequently implemented functional and flat organizational structures, you often find companies that pursue what they will refer to as a matrix design. In its purest sense, the matrix structure seeks to address the needs of the organization based on two unique business perspectives as exhibited in *Figure 11*. Think of a spreadsheet where columns represent one dimension such as the business functions of sales, marketing, operations, engineering, etc. Now consider the rows to represent the second dimension such as the projects or products the company deems critical. In this spreadsheet-like example the cells represent the individual contributor, or more often a group of contributors, assigned to the initiative based on the function and product/project to be delivered. An example might be representatives of an engineering function fully dedicated to Product XYZ.

By nature of its structure, matrix organizations typically have two chains of command, one responsible for the function and the other responsible for the product or project. It is generally accepted that this type of structure best fits large companies with complex organizational requirements. That said, you will frequently hear many small and medium organizations inaccurately refer to their structure as matrix instead of what will more likely be a team-based organizational design, something I will cover later in this chapter. The simple way to validate

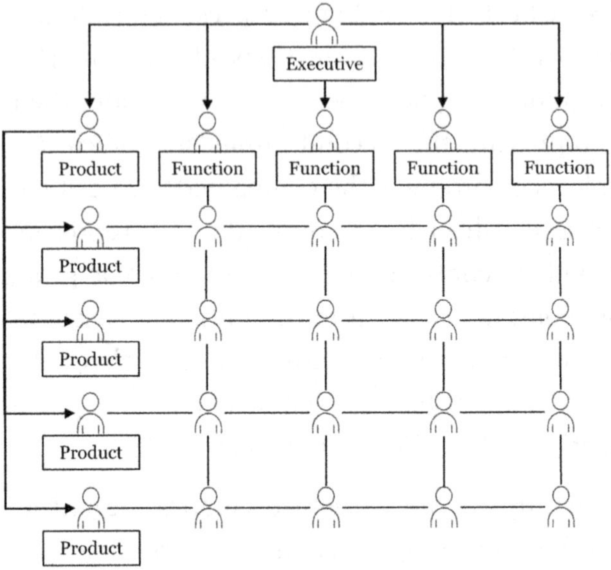

Figure 11

Matrix Organizational Structure Example

that an organizational design is truly matrix is to determine if there are two clear oversight paths for each individual contributor. The challenge for the matrix organization enters the picture when the two chains of command have a different sense of priority. The tendency of the functional manager to selectively move skilled resources between projects to address new or changing demands creates impacts beyond their personal leadership view. Focused Macro Context discipline is a necessity to ensure management does not shift focus to lower priority projects without proper synchronization, which becomes tempting with a finite set of resources pursuing specific delivery demands.

Suffice it to say that alignment in the matrix structure is extraordinarily complex across an entire business entity. Each resource is highly focused on its specific functional and product/project responsibilities, and management inherently views the use of the associated resources differently. Within the matrix structure, alignment requires that the management of each function and product/project element be keenly aware of the priorities of the other. To reiterate a key premise, consistent and frequent Macro Context communication is critical to success in

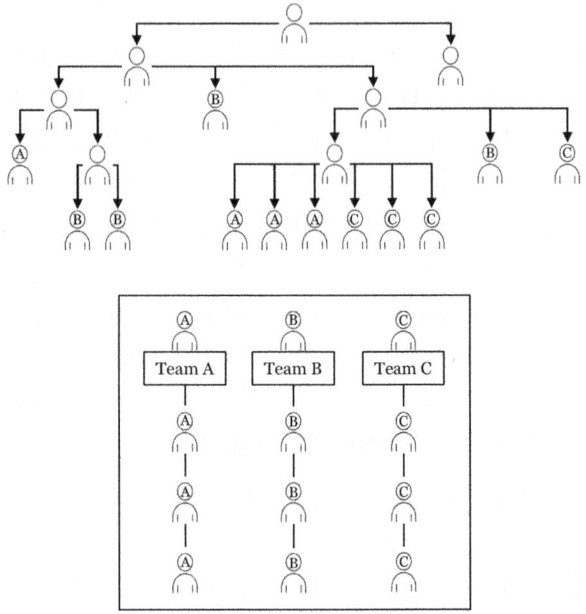

Figure 12

Team-based Organizational Structure Example

this structure.

Given that matrix structures are most often applied in large companies to address complex organizational needs, it is difficult to apply a single standard for alignment that would meet all conditions. Alignment in the matrix structure requires a multifaceted assessment, so I am not quick to conclude that a simplified solution such as *The Alignment Prescription* will automatically fit the need. The A5 Principle involving awareness, accountability, and achievement will apply, however the mechanism for achieving it may not be the same as with other organizational structures.

Now let us look at a fourth type of organizational structure, one that has become increasingly popular. Team-based structures are frequently leveraged based on the ability to implement them in conjunction with other organizational designs. For example, it is not unusual to find a team-based structure integrated with a functional organization. *Figure 12* exhibits this type of condition where the organization is structured functionally, while individual contributors are assigned to specific teams. Keep in mind that this example is one of many ways that team-based structures are implemented.

Many smaller and mid-size companies confound team-based structures as a matrix design, but they are significantly different. In team-based structures, individual contributors are assigned to a team due to specific skills and experiences, typically for the duration of a project or initiative. In some companies, teams are created more permanently and not immediately disbanded at the completion of a specific project or initiative, but those are primarily conditions with a significant level of consistency in the type of projects that are being pursued. Teams are frequently self-managed, or in some situations a team leader may have responsibility for interacting with management on behalf of the team. In this type of structure it becomes important to ensure there is strong internal alignment at the Macro Context level so that there is a common management perspective of impacts to priorities, projects, and status.

Team-based organizations seek to escape the limitations of hierarchy by providing a mechanism for self-management and empowerment. While in the ideal world, individual contributors would be dedicated to a single team, there are many conditions that exist where an individual contributor would be assigned to two or more teams. Such a situation is among the challenges of team-based structures, particularly given that teams operate at a distance from direct management. But many leaders believe the benefits of communication and empowerment far outweigh the challenges of a team-based design. To that point, you will frequently see team-based structures leveraged in agile environments due to their adaptability and strong fit with the scrum framework.

Team-based structures have a natural linkage for internal communication, so the various contexts tend to be shared fluidly within the team. But outside the direct team, the Macro Context remains significant in its need for active mechanisms to ensure all parties remain aligned around initiatives and direction. While requirements and dependencies tend to be more evolutionary in the agile application of team-based structures, they are not void of the schedule, budget, and resource demands of the Micro Context level. When each member within a team-based structure is providing a consistent flow of information to interested stakeholders, particularly at the Macro and Micro Context levels, the alignment of the team will soar beyond the boundaries of the structure.

Let me make one final point of clarification on organizational design as it relates to exceptionally large companies. Typically these corporations are addressing more complex structural requirements and the need for alignment at the detailed level becomes focused within a specific function/product/project area, an organizational sub-structure in a manner of speaking. Because of this, you will frequently see large companies leveraging hierarchical, horizontal, matrix, or team-based design within their specific divisions. Depending on the strategic focus of the corporate entity, this divisional structure may be defined based on geographic, regional, sectional, modular, or product boundaries. To that point, we will not be addressing the broader alignment between these divisional boundaries within the scope of this book, but instead focus on the internal alignment of their sub-structures, consistent with each of the organizational structures we have discussed.

I began this chapter by explaining that I am not attempting to justify any specific organizational approach. As the CIO, CTO, or senior technical leader, you have assessed the strengths and weaknesses of each structure and determined what best fits your needs. Our goal is to assist you in achieving effective alignment by highlighting the critical points of focus influenced by the structural design of your organization. But the bottom line is that the A5 Principle applies to every structure, internally within your technical organization and externally to your business partner, clients, and customers.

I do believe the emphasis of *The Alignment Prescription* does shift a bit based on the organizational structure in which you are operating. In the functional and horizontal organization, internally you must stay on top of the prescription ingredients from top to bottom, but there should be increased emphasis on the Meso Context. Similarly, in the matrix and team-based organization there should be increased emphasis on the Macro Context, but all contexts are important. Flatter organizations loosen some of the rigidity of the hierarchical structure, and team-based organizations simplify some of the complexity of the matrix structure, but the ingredients to effectively establish internal alignment remain quite consistent. Priorities, projects, status, expectations, requirements, dependencies, schedules, budgets, and resources, all have an important role to play in internal alignment.

The Alignment Prescription contains the critical ingredients to ensure key stakeholders are transparently informed of critical elements of initiatives in most organizational structures. Despite their differences, the constant among various organizational designs is that the context levels play a critical role in building and leveraging an A5 process that seeks to ensure everyone is fully aligned on direction and progress while being provided an appropriate level of detail. In the next chapter we will further explore the process associated with *The Alignment Prescription* and the A5 Principle.

Before continuing, I do want to point out that we have developed an easy-to-use alignment assessment tool for your use. The A5 Alignment Assessment is provided in a short survey format that poses key questions and provides an alignment score for your consideration. It will provide feedback regarding your level of alignment within your organization along with potential areas of improvement. If you are interested in leveraging the assessment tool, it can be accessed at https://executechsolutions.com/assess.

https://executechsolutions.com/assess

"Concentrate on what will produce results rather than on the results."

— Bill Walsh

CHAPTER 7
THE ALIGNMENT DECISION

Through the course of this alignment journey we have been building a simple approach for achieving consistent positive results amidst frequently changing demands. I have identified that the primary focus of this book has been on creating the process for internal alignment within your technical structure that leads to achieving broader business alignment. The goal was to provide a simple remedy for solving the internal challenge with the understanding that the solution will then begin to make its way across broader boundaries. I have highlighted the importance of being on the same page internally, from top to bottom, as a prerequisite to the goal of ultimately creating alignment across the entire business enterprise.

Over the years I have trialed many different solutions geared toward creating internal organizational alignment. One consistent factor to success was to keep the process simple and leverage a variety of mechanisms to foster communication, demonstrate transparency, and build trust with my employees. To that point, for more than a decade I leveraged a monthly communication process that employees referred to as the Birthday Meeting. Simply stated, based on their month of birth, we invited our technical employees off site for an informational business lunch session, providing them insight into the projects the management team was pursuing. It was a different facet of building internal alignment, but it was fully consistent with the alignment goals for establishing an environment of communication, transparency, and trust. By restricting participation to their birthday month, roughly six to ten percent of the employee population participated in each session. Our focus at the meetings was to discuss the top priorities being pursued along with any number of other topics we felt were important to address. A typical two-hour session would involve the interactive participation of between fifteen and twenty participants, personalized through introductions and guided commentary. There was always plenty of opportunity to ask questions on the subject matter presented, or anything else for which an employee had interest. Other than specific personal human resources related questions, we had no limits on the discussions.

One of my former senior leaders recently commented that one of the beauties of the birthday month selection approach was that we brought together employees who did not necessarily interact during normal business. It opened communication channels among employees from different parts of the organization and created comradery among varied teams. It allowed for the development of new relationships, even if it was as simple as creating hallway hellos, and it synchronized an understanding of what was important (WIN).

I mention these birthday meetings for a couple of reasons. For one, if your organization is of the size that warrants an extended communication channel, it is certainly an approach worth some consideration.

Understand that this approach does not meet all needs and, while customizable, the success is somewhat dependent on your leadership style. More importantly, I have emphasized communication and trust as critical prerequisites in creating alignment and these employee sessions were all about validating a commitment to employees in that direction. While this approach may be a bit of a tangent to the specific contexts of *The Alignment Prescription*, I provide it as food for thought as it relates to addressing the broader communication needs of your organization.

One of the primary goals of this book was to provide new avenues you can pursue to get a jump start on addressing the alignment dilemma. As the CIO, CTO, or senior technology leader for your organization, you have reached a point in this discussion where you are able to make an alignment decision. You can choose to maintain the status quo, but let us be honest, that is not why you are reading this book. You can choose to evaluate a more complex company-wide alignment initiative, as organizations frequently do, although that still leaves you with the need to address your internal alignment requirements for the broader initiative to succeed. Or you can choose to implement some or all aspects of this proven remedy to take you beyond the status quo, internally align your technology structure, and pave the way to the broader business alignment goal. Let us explore those options a bit.

Advancing the alignment state of your organization no longer needs to be thought of as lengthy, complex, or requiring extensive business coordination. By leveraging *The Alignment Prescription* methodology in your technical organization, no matter the size, you now have the capability to launch an approach that will create instant project visibility and accountability among your employees. If you choose to pursue the approach aggressively, you will be able to immediately improve the flow of information, consistency of communication, and transparency of progress, or lack thereof. When you focus on the elements of the A5 Principle, you will establish an approach that creates awareness, ensures accountability, drives achievement, and solidifies internal alignment. You will empower your organization by placing critical information in their direct control. Without a lot of fanfare or disruption, you will have bolstered the achievement of consistent positive outcomes within your organization, the lack of which being the cancer we set out to cure.

Now there are varying degrees at which *The Alignment Prescription* may be pursued. At its highest level, internal alignment requires you and your organization to intensify its focus on priorities. Going back to the earlier quote by Jim Barksdale, "the main thing is to keep the Main Thing the main thing." I might put a plural on the Main Thing based on typical technology organization workloads, but a priority focus is the most meaningful way to ensure your organization is internally aligned. If you walk away from *The Alignment Prescription* with only one thought, let it be that effectively managing priorities is critical to the success of your organization. I encourage you to engage yourself and your employees in a process of identifying the projects that are critically important and make certain that at any point in time you understand the state of their progress. To be clear, this is not simply a mental thought process of what is important, this is a documented statement that you and your employees are able to share. This is by far the most essential ingredient of *The Alignment Prescription,* and it is a key driver of the A5 Principle. It is a step into the Macro Context in lieu of fully engaging in the A5 process.

Using this approach, your vision of what is important, as compared to how others interpret that vision, must be reviewable. Whatever method you choose, select a communication vehicle for making your priorities visible, establish a timeframe for periodic updates, and invite your organization to review them. *Figure 13* is one example of a priority list you might leverage, or simply create your own version. If you have any hesitancy in implementing a robust prioritization process, then begin with yourself and share the information with your direct team. When you feel ready to engage your leaders in the process, do not hesitate to expand it. And to be clear, this is an ordered priority list, not a single or random list of organizational directives. We all know that juggling the priority list is a persistent challenge that many technology organizations face, so create your Top 10, or whatever number of projects you feel is appropriate. The point is to make certain that you and your organization always know what initiatives are most important and how they are progressing. If this is the only step you take after reading this book, you made a significant move toward being internally aligned.

If you prefer to pursue *The Alignment Prescription*

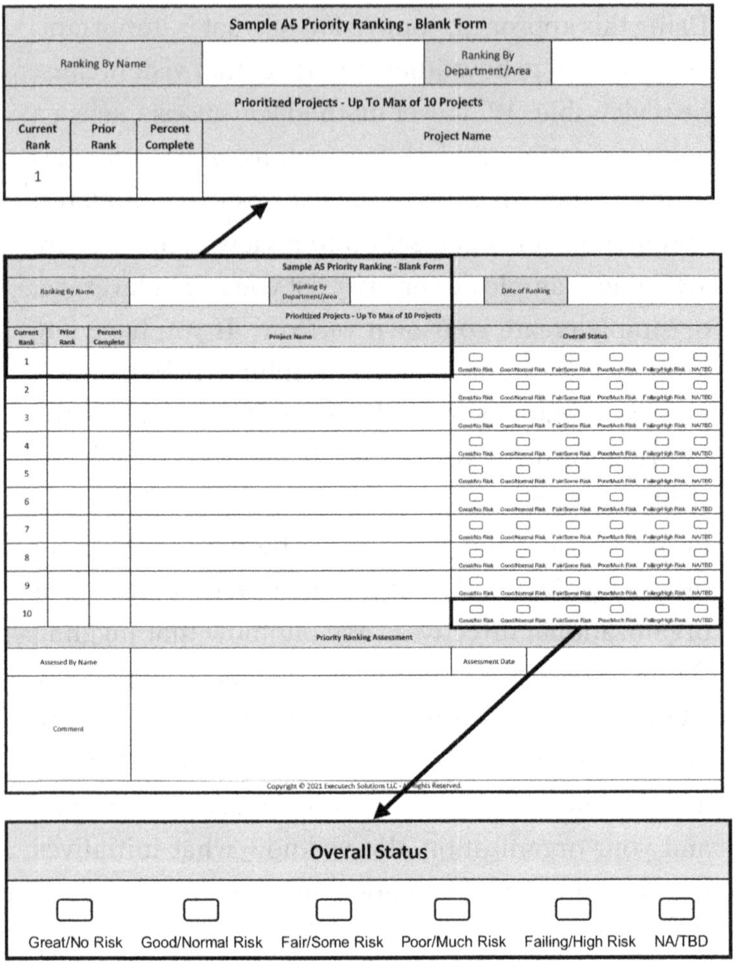

Sample A5 Priority Ranking - Blank Form			
Ranking By Name		Ranking By Department/Area	
Prioritized Projects - Up To Max of 10 Projects			
Current Rank	Prior Rank	Percent Complete	Project Name
1			

Sample A5 Priority Ranking - Blank Form					
Ranking By Name		Ranking By Department/Area		Date of Ranking	
Prioritized Projects - Up To Max of 10 Projects					
Current Rank	Prior Rank	Percent Complete	Project Name	Overall Status	
1				Great/No Risk Good/Normal Risk Fair/Some Risk Poor/Much Risk Failing/High Risk NA/TBD	
2				Great/No Risk Good/Normal Risk Fair/Some Risk Poor/Much Risk Failing/High Risk NA/TBD	
3				Great/No Risk Good/Normal Risk Fair/Some Risk Poor/Much Risk Failing/High Risk NA/TBD	
4				Great/No Risk Good/Normal Risk Fair/Some Risk Poor/Much Risk Failing/High Risk NA/TBD	
5				Great/No Risk Good/Normal Risk Fair/Some Risk Poor/Much Risk Failing/High Risk NA/TBD	
6				Great/No Risk Good/Normal Risk Fair/Some Risk Poor/Much Risk Failing/High Risk NA/TBD	
7				Great/No Risk Good/Normal Risk Fair/Some Risk Poor/Much Risk Failing/High Risk NA/TBD	
8				Great/No Risk Good/Normal Risk Fair/Some Risk Poor/Much Risk Failing/High Risk NA/TBD	
9				Great/No Risk Good/Normal Risk Fair/Some Risk Poor/Much Risk Failing/High Risk NA/TBD	
10				Great/No Risk Good/Normal Risk Fair/Some Risk Poor/Much Risk Failing/High Risk NA/TBD	
Priority Ranking Assessment					
Assessed By Name				Assessment Date	
Comment					

Overall Status					
Great/No Risk	Good/Normal Risk	Fair/Some Risk	Poor/Much Risk	Failing/High Risk	NA/TBD

Figure 13

Sample A5 Priority Ranking Report

remedy in a more impactful manner, you will need to engage beyond the priorities of the Macro Context. This leads us to other options for initiating an A5 process in your organization. At one end of this spectrum is an approach among a select few individuals or teams with gradual expansion. At the other end is the more robust implementation across your organizational footprint. Essentially your choices will range from an initial test of the methodology within a specific area, a selective implementation among specific leaders, or the broader initiative that fully engages your organization. Let us explore these options and, logically, you will identify other variations to consider.

Earlier in the book we introduced the A5 Scorecard as a mechanism for communicating progress and changes. While we were focusing on the Meso Context at the time, the Sample A5 Scorecard (*Figure 7)* represented all the key ingredients of *The Alignment Prescription.* Using the A5 Scorecard as a template, an approach for introducing the A5 process would be to leverage the scorecard with select individuals, at a certain management level, or within a specific team in your organization. If you have defined project manager roles, or a project management structure, another option would be to utilize the process among those specific roles. While these trial approaches may have some constraints from the manual processing perspective, if you place limits on the number of initiatives being prioritized and the number of individuals engaged, then it becomes very reasonable to accomplish. This is among the approaches that will allow you to become familiar with the process before pursuing the broader implementation.

As part of any trial or pilot, I do recommend some form of master priority tracking such as the Sample A5 Priority Ranking report previously discussed (refer to *Figure 13*). For a small team you might ask the group leader to manage the ranking report while the members track the periodic progress of the initiatives, weekly recommended. As with the A5 Scorecard reviewed earlier, if you are interested in obtaining a copy of the sample A5 Priority Ranking report, both a blank form and a completed report are at https://executechsolutions.com/download. The bottom line is that no matter how you choose to implement priority tracking, some form of scorecard and ranking approach should be utilized to establish effective monitoring mechanisms.

Clearly, the challenge with a manual priority tracking approach is the overhead it may create as you expand your pilot and the usage begins to grow. The burden of weekly updating of scorecards and ranking reports is reasonable to address on a small scale but expanding participation will create some inefficiencies that will need to be overcome. But with limited participation, or on a trial basis, this does not represent anything insurmountable. Spreadsheet approaches are popular for smaller implementations and are very reasonable to accomplish. If you decide to start with that type of approach, you will want to implement a management process to ensure updates are consistent and accessible. It is one way to trial the priority process, but you will ultimately need to address some form of automation as you expand the usage.

To benefit most from *The Alignment Prescription* process, you ultimately will need consistent use across your full organizational scope. This is the robust option you will need to pursue to gain the most value from the process. As previously mentioned, my company has addressed this requirement with the development of the A5 Prioriteyes software, a product we proudly created to enable technology organizations to deploy *The Alignment Prescription* process quickly and effectively to whatever scope of participation they choose to pursue. More information on the product can be found in the next section but let us first explore what needs to be considered in implementing the A5 process throughout your organizational footprint.

When we first implemented the A5 process in small teams, it was enlightening to see how easily everyone began to engage in it. Internal groups quickly recognized the enormous value of understanding the priorities of others. In the spirit of transparency, initially there were some concerns expressed regarding additional reporting requirements on already full workloads, but once implemented that concern quickly dissipated as the need for other reporting processes were replaced. Depending on your current reporting practices, you may want to consider opportunities for the process to replace or streamline existing approaches for project status reporting. More importantly, the visibility of Macro Context data among individual contributors, as well as their peers, staff, and management, will be quickly recognized as adding significant value.

As hinted at earlier, perhaps even more striking was how business partners took an interest in the A5 process. Within a few months of having fully implemented an early variant of *The Alignment Prescription* product within the technical organization, business partners were inquiring about participating. What accelerated this interest was that we had automated the approach using a tool we referred to as PPS. The Projects, Priorities, and Status (PPS) application was the first generation of an A5 toolset we had developed. There was a recognition by business leaders that by engaging in the use of the PPS tool they would have complete visibility into the priorities of the technical organization and gain an improved grasp on the various initiatives in their own areas. The automation simplified the implementation for business partners and, as a complimentary result, the technical organization equally benefited by having direct access to the priorities of the participating business functions. In the spirit of Lou Holt's What's Important Now (WIN) philosophy, in all parts of the organization there became a clear view into what everyone identified as being most important.

If you are intrigued by the opportunity to implement

an A5 process on a broader scale, I would recommend you begin by leveraging *The Alignment Prescription* as an internal technology focus, then broaden it as a business focus in a follow-up phase when you feel ready to expand its use. Even within your own structure, the simplicity of *The Alignment Prescription* allows you to start to the degree you feel comfortable. Some will choose to start with one or two small teams and grow the process, expanding out like spokes on a wheel, so to speak. Others will choose to implement it at the department level or more broadly throughout the technical organization. Do not let the scope of participation deter you in either direction as it is perfectly reasonable to define it as you see appropriate. Again, begin within your specific scope of control, then later confidently champion the process to other areas. Consider your internal implementation a hook into the broader business alignment you ultimately seek to achieve by allowing the business to observe its value. Most likely they will request to participate as they observe its use. Rather than pushing a solution on those outside your organizational scope, navigate within and allow the process to expand naturally.

Once you make the decision to expand beyond your

scope of control, there will be many aspects of broader business alignment that will still need your attention. As it should be, your initial attention must be focused on internal alignment. That said, technology organizations frequently struggle to get clear visibility into business plans, but by leveraging the transparency of *The Alignment Prescription* you will introduce an approach that will intrigue those outside the technical organization and allow for more of those business priority discussions to take place. As increased awareness of your process begins to develop, you will be approached with interest to participate. This is particularly true with business areas that have the most frequent interaction with your technical organization and feel there is much to gain by participating. By allowing for their involvement in your process, you get visibility into their priorities as well. You will also find that by providing business partners visibility into your priorities, you will promote collaboration based on an improved understanding of the impact of decisions. It creates the environment for two-way participation in planning which will bring the technology organization into earlier discussions with business partners as to what is needed, what is involved in accomplishing it, and what are the appropriate expectations on getting there. To that point, the role of the business partner related to the Meso Context becomes noticeably clearer to them and creates a greater understanding of how the business influences results within the technical structure.

As I have emphasized, the focus of this book has been on resolving the problem that a lack of alignment creates, working from the inside out, beginning with your own organization. As the CIO, CTO, or senior technology leader, the demands on your time have never been greater and the need for efficient solutions are essential. I was committed to providing a solution that was simple, quick, and proven effective.

Implementing *The Alignment Prescription* can be as basic as establishing your own process for prioritization or as advanced as making the decision to leverage our proven alignment product. Whatever solution I presented, it needed to assist you in moving your organization in a direction that would place it in the best possible position to be aligned from top to bottom. It needed to enable you and your organization to achieve consistent positive results, discover how to WIN big, find the cancer cure! It was important that *The Alignment Prescription* provided an approach to build awareness, ensure accountability, drive achievement, and solidify internal alignment. My simple remedy to the alignment challenge is for your organization to be committed to working on the right things at the right time, with the right contexts in focus, knowing that it will lead to the achievement of consistent positive outcomes. No matter how you choose to pursue it, if you embrace the A5 Principle as prescribed in *The Alignment Prescription*, you will take an important step toward remedying the alignment dilemma that is critical to cure.

A5 Prioriteyes Software

A5 PRIORITEYES

We have taken an extensive look at the process built upon the A5 Principle of internal alignment in *The Alignment Prescription.* Related to this, I have identified the software product, A5 Prioriteyes, that is a consideration for fast tracking your alignment implementation. If you have arrived this far, you have developed a strong understanding of the principle and the process, so let us take a few minutes to expand a bit more about the software product.

My company, Executech Solutions LLC, is focused on providing clients with software that aligns internal organizations, drives productivity, and ensures positive results. I do not take this challenge lightly and I am passionate in ensuring that A5 Prioriteyes fosters an effective and consistent management approach amidst a massive number of projects and deliverables.

Our software is built based on the A5 Principle as described in this book. Its design ensures the product aligns with the goals of busy leaders and their core management team. Business decision makers will be impressed with the fact that the software is a low-overhead fully vetted product with proven success. A5 Prioriteyes software has been developed for seamless introduction into corporate culture while bypassing complex integration requirements that often derail major new product implementations.

While you may choose to take the concepts presented in *The Alignment Prescription* and slowly integrate them into your organizational structure, leveraging our software product will dramatically accelerate achieving the benefits of the approach. The approach builds the awareness, the awareness ensures accountability, and the accountability drives achievement. It is an effective and efficient tool to quickly create the internal alignment that drives organizational success. And further down the timeline, our software presents an opportunity to smoothly expand the process, at your discretion, by engaging with your business clients.

From the product implementation perspective, A5 Prioriteyes software leverages a secure SaaS cloud-based design that will allow enterprises to enjoy the benefits of robust capabilities without requiring major investments in new software, hardware, or infrastructure. A5 Prioriteyes software creates visibility into the priorities of others enabling all contributors to experience the benefits of increased awareness, accountability, and achievement that solidifies alignment. We use "eyes" in its name to emphasize the point that with the A5 Prioriteyes software, clients will see an immediate change in their organizational successes. Our software product puts a spark in the traditional way of managing initiatives, ultimately driving productivity, and creating the alignment that many senior leaders seek to achieve. The software product establishes a mechanism for senior leaders and their core management team to develop an understanding of how employees at every level of their organization view what is important. Consistent with the philosophy of legendary coach Lou Holtz, the software assists organizations in focusing on "What's

Important Now" to create alignment and drive positive outcomes.

I do want to point out an important distinction regarding this software. The A5 Prioriteyes software is not a project management tool, it is not focused on task management. There are numerous products in the market that address the tracking of a multitude of tasks and dependencies required in the pursuit of an initiative. The critical need for alignment within an organization requires an important distinction between managing priorities versus projects. As presented in *The Alignment Prescription*, organizational alignment requires a level of coordination across the management team, vertically and horizontally. The details of a project plan often complicate the effective flow of information across an enterprise footprint and requires each manager to make their own assessment of what information is important and what need not be shared.

Projects are different than priorities, and the A5 Prioriteyes software treats them independently, focusing on supplying information at the context level appropriate for a broad range of management participants. Ask a manager about their projects and they will provide you a long list. Ask about their priorities and, after a pause, they will ultimately streamline their response. A5 Prioriteyes software ensures such streamlining is always consistent among all employees at all levels.

A5 Prioriteyes software establishes a consistent approach for enterprise-wide priority management. Implementation may begin with small management teams or more broadly across the corporate footprint. Recognizing the impact of cross-functional dependencies, the software process creates alignment by ensuring synchronized prioritization across organizational boundaries, not simply within a specific project management team. A5 Prioriteyes software creates a steady focus on priorities, tracks progress at an appropriate stakeholder level of detail, and drives project level achievement that frequently translates to both top and bottom-line results. Of critical importance, A5 Prioriteyes software clearly identifies priority ownership and responsibilities ensuring accountability for results among managers, project leaders, and their teams.

We proudly developed the A5 Prioriteyes software to enable organizations to benefit from the A5 Principle quickly and effectively, as defined in *The Alignment Prescription*, based on whatever scope of participation leadership chooses to pursue. If you would like to view an introductory presentation of the A5 Prioriteyes software, simply enter the URL in your browser: https://executechsolutions.com/intro.

https://executechsolutions.com/intro

USEFUL LINKS

THE NEXT STEP

Whether you choose to leverage the principles of *The Alignment Prescription* as informational or directional, I am more than willing to provide any input you require, just link up with me. Should you consider it informational, I hope that the knowledge you have gained will enable you to leverage the A5 Principle in your internal approach toward creating alignment in your organization. To that point, as presented in the beginning of the book, I promise to respond to your questions, and you can reach me directly at https://executechsolutions.com/contact.

https://executechsolutions.com/contact

Should you consider *The Alignment Prescription* to be directional, with or without consideration for leveraging our software solution, I hope that you would consider subscribing to our mailing list. We will periodically provide information about alignment, the A5 Principle, and A5 Prioriteyes, as well as helpful tips and stories to assist you in your alignment journey. We do not sell list information to third parties, and we will not bombard you with spam of any type. Your information is strictly for use by Executech Solutions LLC. We will provide you with useful information for your review on a periodic basis, never more than once weekly. If you are interested in subscribing to our distribution list, you can do so by entering your email address at https://executechsolutions.com/subscribe.

https://executechsolutions.com/subscribe

Finally, if you are interested in learning more about the A5 Prioriteyes software solution we would appreciate the opportunity to discuss it with you. We are confident our software product will quickly demonstrate its effectiveness as a mechanism for generating awareness, accountability, and achievement within your organization. Simply provide your preferred contact information at https://executechsolutions.com/meeting or drop us a line using our info@executechsolutions.com email address, and we will promptly respond.

https://executechsolutions.com/meeting

Throughout this book I have provided other useful links to A5-related information. What follows will provide the specifics for several of those links.

The Sample A5 Project Scorecard and Sample A5 Priority Ranking Report (pdf format) are accessible for downloading. Completed examples of the forms are also available. To access the files for downloading visit https://executechsolutions.com/download.

https://executechsolutions.com/download

The A5 Alignment Assessment tool is provided in an easy-to-use survey format that poses key questions and provides an alignment score for your consideration. To access the informative and enlightening A5 Alignment Assessment tool visit https://executechsolutions.com/assess.

https://executechsolutions.com/assess

To learn more about the A5 Prioriteyes software, you may view an introduction to the software by entering the following URL in your browser: https://executechsolutions.com/intro.

https://executechsolutions.com/intro

Brad Francis, Founder & CEO
Executech Solutions LLC

ABOUT THE AUTHOR

Brad Francis is the Founder and CEO of Executech Solutions LLC, a software product development company that seeks to take traditional thought and transform it by injecting new ideas into its design. His expertise traverses a broad range of experiences including author, technology executive, business consultant, manager, analyst, teacher, and coach.

With over thirty years of information management expertise, Brad has been responsible for all aspects of technology including strategic planning, portfolio management, software engineering, business intelligence, quality assurance, enterprise architecture, networks, data centers, vendor management, and a multitude of technology support services. During his career Brad has been frequently recognized for his contributions toward achieving organizational goals and team-based success. He is particularly proud of having received prestigious leadership awards for bringing together departments to focus on common goals and for introducing new solutions and products that lead to corporate success.

When he is not focused on his clients, Brad is the proud husband of his lovely wife of several decades, father of three, and grandfather of two. Prior to pursuing the business side of technology, Brad was a mathematics and computer science teacher at the secondary school level, as well as head men's basketball and head women's softball coach. A Pennsylvania native and avid sports enthusiast, he currently resides in North Carolina.

Most recently Brad has partnered with his son, Kevin, on the advancement of the Executech Solutions A5 Prioriteyes software. Kevin Francis is a computer scientist with over ten years of experience as a software engineer and project lead. The A5 Prioriteyes software has been designed based on over two decades of Brad's experience in executive technology management and enhanced by leveraging Kevin's expertise in software development, security, and project management. To learn more about the company, be sure to visit the business website at https://executechsolutions.com.

https://executechsolutions.com

To learn more about Brad and his experience, visit his personal website at https://bradfrancis.info.

https://bradfrancis.info

CiO Rx

BONUS OFFER REMINDER

Your purchase of this book entitles you to download **The Alignment Prescription Companion Guide** for **FREE**! This CIO Rx Companion Guide provides the reader with references to each of the book's visuals including *The Alignment Prescription* models, drawings, charts, sample tools, and quotes. The guide is intended to enhance the reading experience by providing quick access to referenced information.

The Companion Guide is only available to those individuals who have purchased, or legally obtained, the book *The Alignment Prescription*. To download your FREE guide, simply enter the URL identified below into your preferred web browser.

DOWNLOAD YOUR FREE GUIDE

https://executechsolutions.com/bonus

CiO Rx

Made in the USA
Monee, IL
04 November 2026

67380798R00085